恐龙的末日

The End of the Dinosaurs

[英] 露丝·欧文/著

刘颖/译

汉英对照
恐龙科普

江苏凤凰美术出版社

全家阅读
小贴士

★ 每天空出大约10分钟来阅读。

★ 找个安静的地方坐下，集中注意力。关掉电视、音乐和手机。

★ 鼓励孩子们自己拿书和翻页。

★ 开始阅读前，先一起看看书里的图画，说说你们看到了什么。

★ 如果遇到不认识的单词，先问问孩子们首字母如何发音，再带着他们读完整句话。

★ 很多时候，通过首字母发音并听完整句话，孩子们就能猜出单词的意思。书里的图画也能起到提示的作用。

最重要的是，感受一起阅读的乐趣吧！

扫码听本书英文

Tips for Reading Together

• Set aside about 10 minutes each day for reading.

• Find a quiet place to sit with no distractions. Turn off the TV, music and screens.

• Encourage the child to hold the book and turn the pages.

• Before reading begins, look at the pictures together and talk about what you see.

• If the child gets stuck on a word, ask them what sound the first letter makes. Then, you read to the end of the sentence.

• Often by knowing the first sound and hearing the rest of the sentence, the child will be able to figure out the unknown word. Looking at the pictures can help, too.

Above all enjoy the time together and make reading fun!

Contents 目录

许多恐龙
Lots of Dinosaurs

恐龙生活在数千万年前的地球上。
有些恐龙是食草动物。
其他恐龙是食肉动物。

Dinosaurs lived on Earth tens of
millions of years ago.
Some dinosaurs were plant-eaters
(herbivores).
Other dinosaurs were meat-eaters
(carnivores).

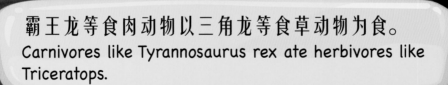

霸王龙等食肉动物以三角龙等食草动物为食。
Carnivores like Tyrannosaurus rex ate herbivores like
Triceratops.

霸王龙 **Tyrannosaurus rex**
(tie-RAN-oh-SAW-rus rex)

三角龙 **Triceratops**
(try-SERA-tops)

5

永远消失
Gone Forever

但现在没有恐龙了。

它们怎么了？

But there are no dinosaurs around today.

What happened to them?

我们知道恐龙很久以前生活在地球上，
因为人们发现了它们的化石。

We know that dinosaurs lived on Earth long
ago because people find their **fossils**.

三角龙化石
Triceratops fossil

科学家认为他们知道恐龙遭遇了什么。

Scientists think they know what happened to the dinosaurs.

霸王龙化石
Tyrannosaurus rex fossil

来自太空的危险
Danger from Space

大约6600万年前，一颗巨大的小行星在太空中快速飞行。

About 66 million years ago a huge **asteroid** was flying very fast through space.

小行星
asteroid

一天，这颗小行星撞上了地球！

One day, the asteroid crashed into Earth!

这颗撞击地球的小行星的直径约有10千米！

The asteroid that hit Earth was about 10 kilometres wide!

巨浪
Huge Waves

这颗小行星撞上地球后，坠落在了海边。
它砸出了一个大坑。
撞击还掀起了巨浪，瞬间淹没了陆地。
许多恐龙因此丧命。

When the asteroid hit Earth, it landed on the seashore.

It made a huge hole.

It also made huge waves which **flooded** the land.

Many dinosaurs were killed.

海浪比25层楼还高！
The waves were higher
than a 25-storey building!

11

暗无天日
A Dark Land

当小行星撞上地球时，尘埃布满了天空，遮住了太阳。

When the asteroid hit Earth, dust filled the sky. It blocked the light from the Sun.

地球变得又黑又冷。

因为没有阳光，植物停止了生长。

The Earth became cold and dark.

Plants stopped growing because there was no sunlight.

火山
volcano

土地裂开，火山爆发。
The land cracked open and **volcanoes** erupted.

灭绝！  Extinct!

没过多久，更多的恐龙死去。
食草性恐龙因为没有植物可吃而死亡。

Soon, many more dinosaurs died.

The herbivores died because there were no plants to eat.

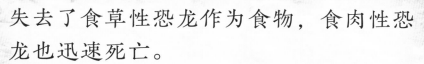

失去了食草性恐龙作为食物，食肉性恐龙也迅速死亡。

The carnivores died because there were no herbivores for them to feed on.

大多数恐龙都灭绝了。
Most kinds of dinosaurs became **extinct**.

巨大的陨石坑
A Huge Crater

科学家发现了小行星撞击地球形成的大坑。

Scientists found the huge hole made by the asteroid.

这处大坑的直径约145千米，
它位于海底，靠近墨西哥。

It is about 145 kilometres wide.
It is under the sea, near Mexico.

小行星撞击而形成的大坑称为陨石坑。
A giant hole made by an asteroid is called
a crater (KRAY-ter).

陨石坑
crater

17

谁是幸存者?
Who Survived?

鳄鱼和恐龙生活在同一时期。

但鳄鱼并没有因小行星撞击地球而灭绝。

Crocodiles were living at the same time as dinosaurs.

But crocodiles were not wiped out by the asteroid.

科学家在霸王龙化石里发现了鳄鱼牙齿的化石。
Scientists found fossil crocodile teeth with T. rex fossils.

鳄鱼
crocodile

19

无处不在的恐龙
Dinosaurs All Around

在小行星撞击地球之前，一些恐龙正在进化成鸟类，它们中的一部分在撞击中幸存了下来。

Before the asteroid crash, some dinosaurs were **evolving** into birds. Some of the birds survived the asteroid crash.

类鸟恐龙
bird-like dinosaurs

它们进化成了今天的鸟类。
They became the birds of today.

鸵鸟 ostrich

鸡 chicken

企鹅 penguin

鸵鸟、企鹅和鸡都是从恐龙进化而来！
Ostriches, penguins and chickens all came from dinosaurs!

21

词汇表 Glossary

小行星　asteroid

宇宙中环绕太阳运动的巨大
岩石天体。
A large rock travelling through space
around the Sun.

进化　evolved

在漫长的时间里一点
一点地发生变化。
Changed bit by bit over a long
period of time.

灭绝　extinct

永远消失。
Gone forever.

被……淹没　flooded

被大雨或上涨的河水漫过。
Covered by lots of water from heavy rains or overflowing rivers.

化石　fossil

存留在岩石中几百万年前的动物和植物的遗体。
The rocky remains of an animal or plant that lived millions of years ago.

火山　volcano

会有燃烧、熔化的岩石喷涌而出的山。
A mountain from which burning, melted rock can erupt.

恐龙小测验 Dinosaur Quiz

① 什么是食草动物？

What are herbivores?

② 什么是食肉动物？

What are carnivores?

③ 当小行星撞击地球时，发生了什么？

What happened when the asteroid hit Earth?

④ 食草动物为何死亡？

Why did the herbivores die?

⑤ 今天的鸟类有何令人惊奇之处？

What is surprising about birds today?

恐龙必备词汇

江苏凤凰美术出版社

prehistoric 史前

A time before people began recording history.

人类开始记录历史前的一段时间。

fossil 化石

The rocky remains of an animal or plant that lived millions of years ago.

存留在岩石中几百万年前的动物和植物的遗体。

scientist 科学家

A person who studies nature and the world.

研究自然和世界的人。

museum 博物馆

A building where interesting objects, such as fossils and art, are studied and displayed.

研究和展出化石和艺术品等有趣物品的场所。

remains 遗体
All or part of a dead body.
尸体的全部或部分。

herd 兽群

A large group of animals that live together.

生活在一起的一大群动物。

hatching 孵化
Breaking out of an egg.
从蛋里破壳而出。

footprint 脚印
A mark in the shape of a foot that a person or
animal makes in or
on a surface.
人或动物的脚
（足）踏过后
留下的痕迹。

reptiles 爬行动物

Animals including dinosaurs, pterosaurs and modern animals such as lizards and alligators.

包括恐龙、翼龙和现代的蜥蜴、短吻鳄等动物。

herbivore 食草动物

An animal that only eats plants.

只吃植物的动物。

prey 猎物

Animals that are hunted and eaten by other animals.

被其他动物猎
食的动物。

volcano 火山

A mountain from which burning, melt-ed rock can erupt.

会有燃烧、熔化的岩石喷涌而出的山。

asteroid 小行星

A large rock travelling through space around the Sun.

宇宙中环绕太阳运动的巨大岩石天体。

extinct 灭绝
Gone forever.
永远消失。

evolved 进化

Changed bit by bit over a long period of time.

在漫长的时间里一点一点地发生变化。

Triceratops 三角龙
(try-SERA-tops)

Tyrannosaurus rex
霸王龙
(tie-RAN-oh-SAW-rus rex)

Troodon 伤齿龙
(TRO-uh-don)

Pterosaur
翼龙

Velociraptor 迅猛龙
(vel-OS-ee-rap-tor)

Stegosaurus 剑龙
(STEG-oh-SAW-rus)

Diplodocus 梁龙
(di-PLOD-u-kuss)

Spinosaurus 棘龙
(SPY-no-SAW-rus)